Responding to people's lives

DEVELOPING ADULT TEACHING AND LEARNING: PRACTITIONER GUIDES

Yvon Appleby and David Barton

National Institute of Adult Continuing Education
(England and Wales)

21 De Montfort Street
Leicester LE1 7GE

Company registration no. 2603322
Charity registration no. 1002775

NIACE has a broad remit to promote lifelong learning opportunities for adults.
NIACE works to develop increased participation in education and training,
particularly for those who do not have easy access because of class, gender, age,
race, language and culture, learning difficulties or disabilities, or insufficient
financial resources.

For a full catalogue of all NIACE's publications visit
www.niace.org.uk/publications

Cataloguing in Publications Data
A CIP record for this title is available from the British Library

ISBN 978-1-86201-330-8

Cover design by Creative by Design Limited, Paisley
Designed and typeset by Creative by Design Limited, Paisley
Printed and

Developing adult teaching and learning: Practitioner guides

This is one of two practitioner guides arising from the longitudinal study of Adult Learners' Lives, carried out at Lancaster University from 2002 to 2005. This study was part of the National Research and Development Centre for Adult Literacy and Numeracy (NRDC)'s programme of research into economic development and social inclusion. The other practitioner guide arising from this study is entitled *Bridges into learning for adults who find provision hard to reach*, by Yvon Appleby.

These guides illuminate the value of a 'social practices' approach to adult teaching and learning, whereby developing skills is part of a wider, holistic way for teachers and other professionals to make learning and achievement relevant and meaningful to learners' everyday lives. They offer practical ideas about how to work with adults in a variety of settings, taking into account their life patterns, circumstances, future plans and hopes.

For more information on the Adult Learners' Lives project, please see
www.nrdc.org.uk

Contents

Acknowledgements   vi

Introduction   vii
   Who this guide is for   vii
   How the guide is organised   viii

1: Background to a social practice approach   1
   Where we are now – what went before?   1
   Adult Learners' Lives   2

2: Teaching using a social practice approach   5
   Social practice principles in action   6
      Example 1 – Everyday uses of numeracy   7
      Example 2 – Student inquiry   9
      Example 3 – Working with young people   12
      Example 4 – Learning grammar through speaking and listening   13
      Example 5 – The notebook: From class to everyday life   16
      Example 6 – Individual work with a volunteer   18
      Example 7 – Individual needs while working in a group   20
      Example 8 – Finding safe routes through learning   22
      Example 9 – Developing speaking and listening through stories and games   24
      Example 10 – The Somali House workshop: Collaborative ways of working   26

3: Challenges and implications   27
   What does this approach bring?   27
   What are the challenges and implications?   31
      Gathering information about people's lives   31
      Using and developing materials   32
      A whole people approach   33
      Professional tensions   35
   In conclusion   36

References   38

Further reading   41

Resources and websites   43

Acknowledgements

Many people have contributed to the fieldwork, research collaborations and ideas represented here. We would like to thank our colleagues on the Adult Learners' Lives project, Rachel Hodge, Roz Ivanič and Karin Tusting, who were closely involved in the original project on which this work is based. Also practitioner researchers Dianne Beck, Gill Burgess, Kath Gilbert, Russ Hodson, Russell Hudson and Carol Woods who worked with us. We appreciate input from members of the advisory group and especially to tutors, managers and trainers from Cumbria and Lancashire for their insightful contributions and critical reading. In particular we would like to thank Ron Creer, Paul Dickinson, Hilary Farrer, Judi Jameson, Norma McCaskill, Meriel Lobley, Sue Ogden and Maureen Stevens.

This work was carried out at the Literacy Research Centre at Lancaster University. Yvon Appleby is now at the University of Central Lancashire.

Peer review

This guide was peer reviewed by:

- John Callaghan, University of Leeds
- Jan Eldred, NIACE
- Carol Taylor, Basic Skills Agency at NIACE

Introduction

In this guide we introduce a social practice approach to teaching literacy, language and numeracy (LLN) to adults. This approach to teaching and learning means taking into account and making best use of the ways in which people use LLN in their everyday lives, thus supporting learners' practices, goals and purposes. There exists a significant body of work about LLN as social practice. But, as yet, there has been little work to connect these insights and evidence with actual practice in adult learning. This guide, written for teachers and others working directly with learners, begins to do this. We draw on theory and research and have worked in collaboration with many practitioners who have contributed ideas, knowledge and experience. The guide aims to help practitioners to think about what literacy, language and numeracy are, how they are used in daily life and what they mean to adult learners – and consider how to develop teaching and learning strategies in the light of their reflections. Teaching and learning can take place in many contexts: college classrooms and workshops, in the workplace, in community settings or within the family. We have no simple answers to provide, but we raise issues and ideas to encourage reflection, stimulate discussion and enable practitioners to develop and use the social practice approach in teaching strategies and practical learning activities.

Who is this guide for

As the field of LLN grows, many more people are training to become subject teachers and support workers. Others, who have worked in the sector for many years, are now undertaking new qualifications and further professional development and training. The principles in this guide can equally underpin initial teacher training and professional development – teachers are also learners, learning and practising literacy, numeracy and language throughout their lives. In addition, as LLN is increasingly embedded in other provision, vocational or recreational course teachers are taking on new knowledge about how best to support their learners' needs. This presents two challenges. First, LLN teachers need to know how, for example, to contextualise materials within vocational and workplace learning programmes; and second, vocational tutors need to know how best to embed LLN learning within their subject teaching.

Although work on social practice approaches has primarily been associated with literacy, a social practice view is equally relevant to numeracy and language teaching. In this guide there are examples from language classes, literacy classes and numeracy classes. Our hope is that practitioners in all of these subjects will find the examples useful. For more on English for Speakers of Other Languages (ESOL), see Roberts *et al.* (2007). For more on numeracy, see Coben *et al.* (2007).

A social practice approach offers an important dimension to teacher education and professional development. It links learners, as people leading complex lives as adults, to teaching. The ideas outlined in this guide may not be new to good practice, or to policy. Here, they have been developed into principles and practice which are based on extensive recent NRDC research, in particular the Adult Learners' Lives project. This project was rooted in practice, exploring adult learners' lives to gain a deep understanding of their learning needs and the practices which would support them. It has helped to develop existing practitioner knowledge, experience and wisdom, both explicit and intuitive, and turn this into evidence which can then support the further development of teaching and learning.

The focus on LLN learners' lives and needs has recently been given new impetus through the policy of personalisation, or the tailoring of learning programmes to meet individual learners' needs, interests and aptitudes – with the aim of motivating learners to engage, persist and progress in learning which is relevant to their needs, life patterns and purposes. The personalisation agenda is set out in the recent Further Education White Paper (DfES, 2006); it gives a strong policy context and new impetus to support social practice approaches.

How the guide is organised

The main purpose of the guide is to offer clear examples of how the principles work in practice. To make the guide accessible, we use general examples and then raise points to reflect upon, helping to anchor the ideas within LLN teaching and learning. To make the guide more useful, we have added an annotated reference section with details about how to find out more.

In the first section, we explain how a social practice view provides ways of looking at the meanings and uses of LLN in people's lives: the things people are bringing to learning, which are part of their identities and are relevant to why they have joined courses. LLN as a social practice will include aspects of people's lives, the places they live and work in and relate to, their networks and events at home, in the community, in leisure and work. Learners use LLN in many contexts: from the local environment – roads and streets, their children's and grandchildren's schools, shops, offices and organisations in towns and villages – to the wider national and international connections that are part of adults' lives. Places, people, and activities have multiple meanings involving multiple practices.

In the main part of this guide, Section 2 – 'Teaching using a social practice approach', we describe five principles that provide a social practice framework for adult LLN teaching and learning. These principles are not a simple template to be applied. Rather, they provide a way of thinking about the complexities which exist in adult teaching and learning environments and help practitioners to consider the power relations in learning environments as well as what both learners and teachers bring to learning. A social practice view enables us to recognise that some literacies, languages and numeracies hold power, official recognition and status, whereas others are considered inferior, informal, marginal or 'non-standard'. In the principles, we see people as active agents in learning, rather than either 'receivers' or 'providers' of the knowledge and skills that support an enhanced ability to practise LLN. Each principle is illustrated with two examples taken from a variety of learning contexts. The examples are drawn from the NRDC Adult Learners' Lives research project[1] which explored in depth the purposes, uses and meanings of LLN by adult learners in daily life as well as in the learning environments. Each example raises different issues and challenges, and provides the basis for further reflection, discussion and possibilities for uses and activities in teaching.

In the final section, we consider what is offered by a teaching strategy which uses a social practice approach to LLN and raise some key questions. We ask whether and how this approach is different and if it can make a difference to learning. The social aspects of learning and literacy are emphasised, and we ask: how does this blend with the development of skills, individual and group attainment, assessment and testing? The social practice approach encourages learners and teachers to ask questions, explore problems and tasks, and seek outcomes and answers for themselves. What are the implications and benefits of this way of working? We ask why and how reflective practice and practitioners engaging in research and development are important in teacher education, training and continuing professional development. Lastly, we provide a list of websites and further reading to help people explore the approach further.

---

[1] See Barton *et al.* (2006), Ivanič (ed.) (2004) and Ivanič *et al.* (2006). More information on this project can be found at: **http://www.nrdc.org.uk/projects_details.asp?ProjectID=14**

1 Background to a social practice approach

A social practice approach does not assume that there is one meaning of literacy, although it recognises that powerful meanings do exist. Literacy as a social practice takes account of the differences in people's lives, including their culture, language, class and status in society. It does not view all adult literacy learners as belonging in a simple, uniform category. Instead it sees people with many different backgrounds, many different current circumstances and many different existing resources and hopes for the future. People bring their own backgrounds into learning, and have their own aims and purposes.

A social practice view recognises how people learn and use what they learn in their lives and communities. This could be searching for a number in a telephone directory, reading junk mail, writing a shopping list or a letter of complaint. It could also be composing a poem, writing a letter for a friend or reading instructions to operate a DVD player. It recognises that literacy is all around us: in signs, in text messages, on computers and is written, spoken and read in many different languages.

One study which used this approach looked in detail at literacy and numeracy in the everyday lives of people in a town in north-west England. *Local Literacies* (Barton and Hamilton, 1998) observed and recorded the variety of ways that people used literacy in their hobbies, in work, for family organisation and in networks and community groups. They found that people employed different types of literacies in different places, for different purposes – from shopping lists to taking notes from the allotment gardens meetings. People's uses of literacy and numeracy were often linked to their 'ruling passions' and were always set within their social and cultural networks.

Where we are now – what went before?

In the 1970s, many people in the UK became acutely aware that, in spite of a compulsory secondary school system, many adults still identified problems with reading and writing. This prompted a national campaign to promote adult literacy and to tackle the 'problem' of illiteracy. Between 1970 and 2000 there were many initiatives and responses to what was termed adult basic education needs, but no national policy (see Hamilton and Hillier, 2006). The Skills for Life strategy, the national strategy for adult language, literacy and numeracy, began in England in 2001 in response to recommendations from the Moser Report (DfEE, 1999). This national strategy introduced standards, a core curriculum, targets, assessment and far-reaching reforms to teacher education.

Many people have been positive about the increased resources, increased professionalism and increased quality that the strategy has created. At the same time there is the danger that focusing on what people lack can mean that the skills they do have are not recognised and that individuals experience blame for what are primarily social problems. While there are connections between social inclusion and low levels of literacy and numeracy, other social factors like poverty, bad housing and unemployment also need to be taken into account. To challenge a 'deficit' model, which sees people as failures, it is important to support learning within the classroom by recognising and developing the skills and knowledge that learners bring from outside it.

Adult Learners' Lives

The material for this guide comes from the Adult Learners' Lives project which ran between 2002–5 (see Note[1], page ix). This NRDC study was conducted by the Literacy Research Centre at Lancaster University. We worked in Blackburn, Liverpool and Lancaster in the north west of England to explore what was happening in adult literacy, numeracy and ESOL classes. We observed classes, worked with learners and teachers and talked to people about how what they were learning related to their everyday lives. In all, more than 282 people participated in the study. Fifty-three learners were followed to explore more detail in more depth over time.

In this research we looked at adults learning in several classes and also examined the learning in the lives of some of those adults. We worked closely with many practitioners who shared their observations and reflections about teaching and learning. Some worked even more closely with us as they developed their own research on issues arising from their teaching practice. We found that a social practice approach can support practitioners to connect what people learn in their everyday lives with the more structured learning offered in the core curriculum. This connection enables a greater understanding of people's motivations for attending a particular class. It also sheds light on what types of things engage people and help them persist with their learning.

Our research method drew upon a tradition that sees people as situated within the culture and society they inhabit. This enabled us to look closely at the types of languages, texts and writing that people used in a variety of settings, for different purposes. We used different methods for gathering information in a variety of forms, such as photographs, video, email, telephone/mobiles, interviews, observations, diaries, visits and conversations. We worked collaboratively with many practitioners and learners, exploring ethical ways of conducting research. We were particularly concerned that the people we worked with were included and represented in all aspects of the research: their voices are often not heard in official discourses about them.

Participants in the research came from many different social and economic backgrounds as well as different language communities. Some had jobs while others were unemployed. Some lived in extended families and communities, some lived on their own, whilst others were trying to build up networks in new surroundings. People had different experiences that shaped their uses and meanings of language, literacy and numeracy as well as their learning. Some had access to foundation education, some did not and others had dropped out. Some refugees, asylum-seekers and immigrants, particularly women, had not had an opportunity to learn written language and were learning and using several different languages in the class, at home and at school. Some came to brush up what they had previously learned, for example spelling or maths, whilst others wanted to learn to prepare for a test or exam. Others were highly educated and learning English as an additional language. People managed different difficult circumstances including poverty, bad housing and homelessness, health issues, violence and fear. They also managed positive experiences within families, communities and work.

Library photo, posed by model, ©*www.johnbirdsall.co.uk*

Although there were specific differences relating to the subject area, type and location of classes, we found many common threads. These were:

- Good personal relationships are important.
- Listening to learners is important, and takes time.
- Learning is based upon what both learners and teachers bring with them.
- Language, literacy and numeracy learning take place outside of the classroom.
- Teachers need to respond to specific contexts and individual learners' lives.
- Teachers need time to adapt to change and develop new ideas.
- Teachers experience tension in their professional identity.

These threads form the basis for developing the approach to teaching and learning which we refer to as a social practice pedagogy.

2 | Teaching using a social practice approach

In this section we describe five principles of social practice pedagogy. They are designed to support thinking about the complexities of adult language, literacy and numeracy teaching and learning. We believe that teaching and learning amount to more than simply 'transmitting' or 'broadcasting' information and knowledge. There is interplay between what people think and feel and how they interact socially.

Learning has three dimensions, all operating at the same time. It is a *cognitive* or mental process and involves thinking; it is an *emotional* process and involves feeling; and it is a *social* process and involves doing. Many approaches just focus on one of these aspects but it is important to keep them all in mind. Learning encompasses these three dimensions as people live in social worlds where they think, feel and respond as individuals and as part of groups. (For more on definitions of learning, see Tusting and Barton, 2006; for a discussion of the three aspects of learning, see Illeris, 2002.)

From our research we have identified five principles for using a social practice approach in teaching. This is what we refer to as social practice pedagogy. The five principles of a social practice pedagogy are:

- *Research everyday practices.* Teachers and learners can investigate their changing literacy practices and the learning practices around them.
- *Take account of learners' lives.* People are complex; they have histories, identities, current circumstances and imagined futures. We need to engage with different aspects of their lives in a teaching and learning relationship.
- *Learning by participation.* Using authentic materials, in tasks for real purposes, helps to make links between learning and language, literacy and numeracy in people's everyday lives.
- *Learning in safe, supported contexts.* Recognising and valuing the social aspect of learning, including physical and emotional safety.
- *Locate literacy learning in other forms of meaning-making.* Recognising and working with different literacies that include oral, visual, individual and group ways of communicating.

Social practice principles in action

Here we look at how these principles can be applied in teaching and learning. We use examples from research and from practitioners to describe different practices in real settings and in different learning environments. The settings, the teachers and the learners, who are all real people being represented by their chosen pseudonyms, come from different areas of practice that include literacy, numeracy and ESOL provision. We have chosen them to illustrate the principles in action, not to represent all language, literacy and numeracy learners or all types of provision. They provide examples of meanings from both teachers and learners and interactions within classrooms that relate to learners' lives outside the classroom. Sometimes this can be seen as an immediate relationship, what the learner is currently doing; at other times it draws upon the learner's past experience.

We describe what we mean by each principle, explaining why it is an important element of the social practice approach. The examples do not necessarily illustrate best practice. They often show 'things in progress', or working within classroom constraints – in other words, mirroring what classrooms are actually like, with insufficient time and competing priorities.

> **PRINCIPLE 1: Research everyday practices. Teachers and learners can investigate the changing literacy practices and the learning practices around them.**

Once we start to question what literacy is, and what part it has in people's lives, we need to develop ways of exploring different meanings and uses. Investigation through inquiry is a valuable way to do this as it enables us to look at things in a more focused and systematic way. It is also something that many of us have experience of, as we use methods associated with inquiry to find things out in our daily and professional lives.

Teachers often rely on their professional knowledge to answer questions that arise in their practice. These questions might include: Is individual work more effective than group work in my spelling class; or, Is working on an individual learning plan the best use of time in the classroom? Learners use a similar inquiry-based system, asking questions like: Will this course give me what I need to help my children with their homework? Both learners and teachers use methods of inquiry inside and outside the classroom as part of their everyday practices. This method can be developed to investigate both literacy and learning practices inside and outside the classroom, drawing upon the experience and insight of both the learner and teacher. The two examples that follow show this.

In the first, maths teacher Kay used a simple questionnaire to understand how people used numeracy in their everyday lives. Her reflection on teaching maths inside the classroom showed she did not have a way of finding out how her students used numeracy outside of it. Her inquiry, while overall beneficial to her practice as a teacher, produced tension in the short term for her between spending time investigating and teaching in the classroom. The second example shows how Jason, a learner from the maths class, investigated some of the literacy practices in his life. By becoming aware of some of his everyday practices, his confidence grew, even though other aspects of his life remained turbulent and affected his participation. He came to see himself as someone who used numeracy frequently, not simply as someone who could not do maths.

Example 1 – Everyday uses of numeracy

Kay, an experienced and well-qualified maths tutor, was concerned that she was unable to find out information about her students' lives and how they used maths in their everyday activities. She used worksheets to enable students to work at different levels, from Entry 1 to support at GCSE level. The worksheets also allowed students to discuss what topic they wanted to cover which was supported through the ILP. Kay therefore had some information about the learners in her drop-in centre but felt her practice would be improved if she could find a way of talking about maths in their everyday lives. This could then be linked to their learning in class.

She undertook a small piece of practitioner research to find this information out using a simple questionnaire that she devised. It asked how people used numeracy in their everyday life, whether for measurement, calculation or other purposes. It also asked how answers were arrived at, whether by asking someone else, guessing roughly or trying to work it out (and if this was the case, by what method). For example:

You need a shelf for an alcove which measures 24 inches. In the shop the shelf lengths are in cm.

Would you:

▇ Have measured in cm in the first place?

▇ Go home and try again?

▇ Ask for help?

▇ Convert it to cm? – please say how

▇ Buy what you think looks right?

Note that these questions are broader than ones that just aim to find out the mathematical result – the questions ask about people's practices. Kay was able to understand her students' uses of numeracy in their lives, which she then related to the maths that she taught in class. For example, she worked with Sarah, a single parent, on measuring for shelves for her children's bedroom – this became the basis of their work on area. Although Sarah's learning was structured around the worksheet, Kay's explanations and applications focused on the context of shelf-building. Sarah was able to translate and use the information from the class in her everyday life. In class she was able to practise using a tape measure and to convert from metric to imperial measurements, as the shops sold lengths of wood using both systems.

©Darren Hester/Shutterstock

Using the questionnaire Kay was able to talk to another student, Jason, about what maths meant to him and how this related to his outside interests. Having found out about Jason's passionate interest in local history she used information on the Mersey tunnel to work on fractions and percentages. She devised questions about the different cost of cars, motorbikes, trucks and vans using the tunnel related to their overall weight. Placing these applications in a context that Jason knew about, and was interested in, made fractions (the application of number) concrete. He found understanding and learning easier and was able to apply this knowledge to other situations.

The questionnaire showed that another student, who worked in a butchery shop, needed extra work on weights and measurements. He was new in the job and did not feel confident in the skills that he used every day. Kay was able to focus his learning on weights, volume and estimation.

A young female student who had come because she said she 'wasn't very good at maths' struggled with multiplication tables, as they seemed difficult and very abstract – 'not very useful' to her and her life. Kay found out that this student was taking a National Vocational Qualification (NVQ) in Childcare and was able to support this by placing multiplication in the context of nappy-changing routines. The student used the nappy-changing record sheet daily but she found it confusing and difficult, and this affected her confidence to succeed in her placement.

Example 2 – Student inquiry

As part of the collaborative research, learners in the maths class took part in an activity investigating what numeracy meant to them in their lives. Several of them took cameras and photographed the use of numeracy in their everyday lives and activities. The images they produced included local shop prices, road signs and information at bus stops. This provided a good tool to discuss what skills people wanted to learn in the class. For example, one learner who sewed realised that she could learn about measurement in class, a skill that would help her to buy the right amount of material from the local market. This was particularly important as she was unemployed and 'money was tight'. Others took pictures showing the view from their college to the River Mersey. This became the basis for work on distance and estimation of the time it takes to make a journey on foot. This was a familiar sight that was turned into a maths question and one that everyone had an opinion about.

A social practice approach enables us to see that things from everyday life are important to support learning in the classroom. It also enables us to consider how skills learned in the classroom are used in everyday life. Jason from the maths class became involved by looking at how he used skills acquired in the maths class in his voluntary work at the tenants' association. He kept a log of his work in the tenants' office, recording his activities and the type of help that people needed. He also recorded wider community activities the association supported on the estate he lived on. Many of these activities used literacy, for example in writing notes, finding and giving information, writing minutes of meetings and making posters advertising community events. They used numeracy in running several bingo clubs, an estate-wide bonus ball (prize draw) scheme, fundraising and keeping the association's books. When other members expressed interest in learning to use the computer, Jason helped the association to introduce various ways of incorporating the use of computers across their work, in the office and in the community. He also helped generate data showing the connection between what people learn in class and what they learn and use in their lives. Although Jason described himself as someone who 'wasn't very good at reading and writing', he operated effectively in an environment that was full of texts, signs and print. The tenants' office contained many notices, sample copies of forms, leaflets, minutes of meetings and notes of work in progress. His experiences, from outside the classroom, show the need to connect this to what he learned inside the classroom.

Library photo, posed by models, ©www.johnbirdsall.co.uk

PRINCIPLE 2: Take account of learners' lives. People are complex: they have histories, identities, current circumstances and imagined futures. We need to engage with different aspects of people in a teaching and learning relationship.

Everyone has a unique combination of experience, interests and needs. Although we talk of groups of people as 'learners' or 'students' it is important to recognise these groups are made up of individuals with different backgrounds, different current circumstances and with different hopes for the future. Some people have stable lives while others have experienced dislocation or difficulties, either as children or as adults. Some are dealing with physical or mental health issues. Others feel that their identity, race, ethnic origin or faith group makes them different from others.

In some areas of targeted provision we may know that people are dealing with alcohol or substance abuse, homelessness, violence or mental health issues. In other areas this may not be visible and yet people may be trying to cope with these same issues in their lives. Taking account of learners' lives means that we need to be aware of what is happening to them and connect this with what they want to learn and how best they are able to learn it. For example, if we know that someone has an addiction and is going to the methadone clinic on Tuesday afternoon, it explains why they are absent from class. If we know that someone is being made homeless, this will account for lapses in concentration and uncompleted homework. We do not need to know everything about learners, and there may be many things they do not want us to know, but we must recognise that there are aspects of each person's life, which, on the one hand, could be drawn upon to enhance learning, or, on the other, may act as substantial barriers to learning.

Many people value the social aspects of the teaching and learning relationship, which recognise them as people with a life outside the classroom. Like any relationship, this is built upon communication and trust. A teacher who understands and works with the complexity of people's lives is more able to link learning in class with life outside it – making learning more positive and meaningful. Where a learner is unable to succeed, perhaps because of external factors, the teacher will be able to limit damage to self-esteem and possibly encourage his or her return to learning in the future. The two examples that follow show this.

The first describes Dee, who, like many other young people, is dealing with the aftermath of bad school experiences and continuing disruption in her life. While her teachers were able to understand some of these issues, they struggled to provide enough opportunity for self-expression and free writing, which they recognised as beneficial to her. There was difficulty in providing different types of learning that Dee needed within a class that focused mainly on outcomes for employment. The second example also shows the tension between teaching knowledge, in this case language, while acknowledging everyday experiences of learners in the classroom. The ESOL class Abdul attended fitted with his expectation of what teaching grammar should be, but he also had to develop his own strategies for learning within it. There was little recognition of the difficulties that he faced as an asylum seeker, in a sometimes-hostile environment, and the impact this might have on his learning.

Example 3 – Working with young people

Dee attends an Entry to Employment (E2E) programme for young people with few qualifications. She is 17 years old and joined the programme because a friend recommended it. The E2E programme is designed to improve young people's basic skills to enable them to become employed. Dee was bullied at school, which affected her confidence and meant she left without qualifications. She likes the intimate, friendly atmosphere of the programme, which she finds different from school and college. She wants to be able to gain literacy and numeracy qualifications, and hopes to get a job as she struggles to live on the £40 a week she receives for attending the programme. While her family are supportive, Dee wants to be independent and find a young people's accommodation scheme.

Despite her initial reservations, she enjoys the DIY and Independent Living modules. With her teachers' help she came to realise that these skills would be useful for future employment and in her everyday life. The Skills for Employment module provided the opportunity to learn problem solving with other learners, including making models, constructing things from metal and wood, loom-weaving and basic plumbing. These practical tasks allowed the teachers to assess wider aspects of learning and fit this with skills needed for potential employment.

While Dee is an enthusiastic learner, the teachers are aware that she is not fulfilling her potential. She has found, and they have recognised, that everyday worries affect her ability to learn. Her mother has been going through a court case relating to the kidnap of Dee's younger sister. Although her teachers saw her as bright enough to 'do anything she put her mind to', they knew that the personal issues she was dealing with were holding her back.

Knowing about some of the issues in Dee's life has meant the teachers have been able to offer her both structure and a space for individual development in the programme. The Skills for Employment module has given Dee the opportunity to learn new skills and a way of assessing how ready she is for work. She has found learning rules and working with others important in realising she has a range of skills required for employment. The challenge for the tutors is to make space for Dee and the others to be able to practise their free writing. The literacy tutor, who recognises the value in expression of ideas, was in a small way able to work in parallel with Dee's own writing at home where she has written poetry and kept in touch with a pen friend in Australia. The tutors could do more for Dee if they could integrate her personal literacy practices into the wider aspects of her learning.

Example 4 – Learning grammar through speaking and listening

Abdul attends an ESOL class that is run four times a week. An agronomist by training, Abdul came from Iraq a year ago and is highly motivated to learn English to be able to get work. He is in his late 40s and lives with his wife and two children. While waiting for his immigration status to be confirmed by the Home Office, he spends his time reading newspapers and books and studying the dictionary to keep up his learning outside of class. He also attends a BBC Learning Centre where, with support, he is learning to use a computer. Abdul has a higher level of written literacy than of speaking, as finding opportunities to practise oral language is difficult. In the class he is particularly keen to learn grammar as it fits with his past educational experiences and helps with language acquisition, keeping him intellectually stimulated.

The tutor, Duncan, uses a variety of methods to teach grammar. On one occasion he cut up a magazine feature and asked learners to assemble it in the correct order. This was then checked against the original copy and learners were asked to identify the type of text it was. Duncan then read the story with very little expression. He used the story as a vehicle for learning grammar and vocabulary, not primarily to interest the learners more widely. He used group comprehension exercises that involved worksheets and encouraged people to feed back answers to the whole group. While he was supportive he did not provide answers directly, encouraging learners to ask each other, or have a go themselves.

Duncan used pair work to recognise verb patterns, asking questions to check the level of individual understanding. He added to this by using different-coloured pens to mark examples on the board for everyone to see. At the end of the lesson he gave out a worksheet to be completed at home, checking that everyone understood what to do and explaining the task using examples from the familiar college setting. Duncan reviewed what had been learned in the grammar lesson and people shouted out what they thought they had learned: 'new words' and 'about adventure, Tony in jungle'. Duncan wrote 'verb patterns', which everyone then copied in their record sheets.

Abdul said he liked that particular lesson because he is interested in grammar and found the task intellectually challenging, though easy to complete. Although Duncan read the story with little or no emphasis, this did not detract from Abdul's enjoyment of it. He found it interesting because it had some adventure in it. In general, Abdul likes Duncan's teaching style because it is 'more detailed and higher level' and 'provides high expectations' for learners. He also likes the way that Duncan asks first before 'telling the answers'.

Abdul sees himself as someone with a strong previous educational background and yet on occasions he has struggled to complete tasks in class. He has found ways of managing this by asking Duncan questions, checking with the people he has been working with and doing additional work at home. Abdul uses both speaking and listening, inside and outside the classroom, to help him learn grammar. The challenge for Duncan was to provide sufficient space for Abdul and the others for more free speaking and writing in order to express their experiences, ideas and current concerns about their new living situation.

> **PRINCIPLE 3: Learning by participation. Using authentic materials, in authentic tasks for real purposes, helps to make links between learning and literacy, language and numeracy in people's everyday lives.**

Most learning occurs in everyday life in real situations and it is often done while participating in activities with other people. Even when learners are working on an individual task, for example a maths worksheet, there is much interaction between people. This includes asking for and offering advice and help. Learning in a social context is about being involved with others as well as being individually engaged.

An important part of encouraging participation is to use materials that relate to people's everyday purposes and activities. This acknowledges that learners have existing skills and knowledge on which to build. Using authentic materials for real purposes encourages participation as people see the value in learning things that are 'really useful' to them rather than abstract information or skills. Whether this is a newspaper brought into the class or travel information from the Internet, what is important is the way it is used. A social practice approach recognises the different uses, purposes and meanings of language, literacy and numeracy, acknowledging that they often cross from one domain to another (Appleby and Hamilton, 2006). This understanding supports literacy teaching and learning that makes and develops these connections. An example of this is working with a learner to write a letter of complaint to the local newspaper about the intended closure of the market. If this is something the learner feels strongly about and has significance in his or her life, it makes materials authentic.

Using authentic materials for real tasks makes us question what is authentic for the people who come to learn. We cannot simply rely on existing materials that assume all learners to be the same. Learners are diverse: they speak many

Library photo, posed by models, ©*www.johnbirdsall.co.uk*

languages and dialects, have different work and hobbies, different amounts of money and health and different social networks. Using a crossword may be authentic for one learner but not for another, as the following two examples show. In the first, Fatima's notebook provides her not only with authentic materials in the classroom but a valuable everyday resource outside of it. As the developing content of the notebooks was discussed with each learner, they were based upon and reflected their everyday practices. As part of this process, it is important to recognise and support learning that is not only functional but fun, challenging and inspirational, and mirrors everyday life.

The second example shows Jack, a dairy farmer, working with a volunteer to link his learning with his farming life. The use of materials based upon his experience gave Jack a sense of safety in the classroom. Their relevance to his everyday life was motivating. Jack had been attending the class for several years and he felt comfortable with the pace and content of his learning, making progress at a pace he could manage.

Example 5 – The notebook: From class to everyday life

Fatima attends a Bilingual Family Learning Centre regularly to improve her English. She was born in the mountainous part of Yemen and as a child had little access to education. Like many Yemeni girls of her generation she stayed at home and worked on the land, contributing to the family economy. Although she speaks Arabic, Fatima is not able to write using Arabic script and does not have time to attend classes in the local Arabic school as she has six children. This means that she is unable to write reminders in Arabic of what she is learning in English. The bilingual support worker commented:

> It's a problem, that's why she forgets easily... She easily forgets the meaning of the word. She reads it again and again and asks, "What's the meaning of this word again?"

Since attending the centre Fatima's confidence in speaking has improved. She is able to ask people to speak more slowly so that she can understand. She practises her speaking with her neighbours, at her children's school and even at the hospital where one of her children had many visits. Writing is more difficult. Even with practice and repetition in the classroom there are important pieces of information that Fatima needs to be able to write in her everyday life that she finds difficult to remember.

The centre devised a simple notebook as a way of enabling learning while building up useful information that could be applied to everyday tasks and purposes. The notebook is a small blank book with a reference page and pages that can be glued in as they are completed. The pages cover personal information such as name, address, children's names, and important telephone numbers. There are also subject groups like Hospital, School, Pregnancy and Time which contain relevant vocabulary and language structure. The books are individually filled in at a pace the learner is comfortable with. The notebook acts as a learning tool, which is in the control of the learner, who decides what is relevant for their everyday tasks and purposes. This is an important example of something designed in class that was found to be useful in everyday life.

The women at the centre found the notebooks gave them increased independence as the information they contained enabled them to fill in forms, make phone calls or contact schools. People reported using them when visiting the doctor, the hospital or finding information to make a bus journey. The notebook is thus embedded in the real world of each learner. The language, grammar and expressions used are set within the context in which they are used in practice, so that people can understand and respond appropriately to situations which they need to deal with.

Fatima uses her notebook for personal information such as her children's ages and dates of birth, which she finds difficult to remember and spell. Since she depends on her literacy in English, she relies on the notebook for spellings like 'Yemen' and 'Arabic' when filling in forms. She finds having days and months

©Petro Feketa/Shutterstock

written out helpful so she does not panic, as she knows she has all the information about herself and her family in her notebook. Several learners' notebooks have been replaced as they have worn out through use.

Example 6 – Individual work with a volunteer

Jack is a dairy farmer who has been attending a spelling class for nearly five years. He works with Hannah, a volunteer, who helps him plan what he wants to learn at a pace he is comfortable with. They often talk about Jack's work, the skills that he uses and the issues that affect him. For example, in one audio-recorded session they talked about a legal case where a farmer had shot a burglar. Jack was interested, as he had had intruders staying on his land without permission. Their talk was woven around the learning content of the lesson. Jack frequently practised a sample from the Dolch list of commonly used words, often reading to himself and then out loud, making sentences. Where he made a mistake Hannah offered a correct version, linked to Jack's experience. Jack found this unthreatening and they joked and teased each other in a trusting, friendly way.

Hannah judged the pace carefully, allowing Jack time to absorb new information and consolidate by practising saying it out loud. When they worked on an exercise about prefixes and suffixes Jack started by working on his own for a time quietly, without Hannah intervening. He had to choose beginnings and endings of words from a list on the worksheet. When finished he read them out, and when they were both happy Jack filed them under the correct curriculum reference. Having completed this they worked on a map-reading activity that Hannah had prepared. This encouraged more discussion about Jack's work. This showed he never visited the cattle market in a nearby town anymore because of changes in farming policy, which he described.

Active participation and social conversation enabled Jack to feel comfortable, as Hannah valued his farming knowledge in an educational environment. This was particularly important as Jack had felt a failure at school and was initially apprehensive about learning as an adult. It also provided Hannah with material that related to Jack's everyday life. Hannah was able to identify words that would be useful to him by talking about his farming methods, such as whether he still grew maize or not.

On one occasion, Hannah and Jack discussed his existing computer skills, after which Hannah asked Jack to write five sentences on the computer using words from his everyday world. He asked for help when needed but was able to work independently on the task. This was followed by writing five sentences on the computer using the words he had been practising. This exercise extended his existing skills, as well as generating new words from their discussion about silage.

> **PRINCIPLE 4: Learning in safe, supported contexts. Recognising and valuing the social aspect of learning, including physical and emotional safety.**

People identify a variety of reasons for wanting to learn. Some people talk about improving particular skills, like spelling. For others, it is about broadening the range of practices they participate in: it could be about keeping up with technology, like taking a computer course to help children and grandchildren. Or they may want to get out of the house to meet people, to gain more confidence or learn new social skills. For some people it could be a mixture of all of these. It is important to recognise what people's initial reasons for coming are, and how these may change as a result of attending.

As we have noted, people also bring with them many different experiences that need to be taken into account. For some, a negative experience at school shapes any future learning as it produces feelings of fear, intimidation and panic. Many adult learners describe having been made to look stupid, being bullied and even physically threatened while at school. Some people bring experience of violence, threat and intimidation. Refugees and people seeking asylum may have encountered extreme violence, brutality and displacement. They may also be dealing with current violence, fear and threat associated with racism. Women, and to a lesser extent men, may be dealing with personal or family violence. These experiences may affect regular attendance, the ability to concentrate or the confidence to socially interact with others. It is important to show clearly that each person in a class is valued, will be listened to and treated with respect.

For people living in poverty, in bad housing, with health issues or communicating in a second or third language, it is easy to become socially isolated. Some adults lack the skills and confidence to participate in learning in classes precisely because it is a social activity. It is important to make sure that the social aspects of learning are non-judgmental and include all learners to establish and maintain a positive learning environment. Many learners' experiences remain undisclosed, as they are not recognised as educational issues or classroom concerns by either learners or teachers. But some understanding of what people are dealing with outside the classroom helps to identify ways in which learning can be supported in the classroom.

Two examples illustrate this. The first, from a spelling class, shows how a positive learning environment was essential for Susan to be able to learn. Her previous negative experiences of learning, related to a hearing impairment, left her with spelling difficulties and a lack of confidence. The relationship between Susan and her teacher was key to finding practical strategies to enable her participation and enjoyment in learning. In spite of this, she found group work difficult because of her hearing, and preferred worksheets and paired tasks. Debbie, the teacher, often spent time in addition to allocated teaching hours supporting the independent learning which Susan undertook to help her keep up.

The second example shows how Cheri used similar strategies to attend the spelling class while managing bipolar disorder. For different reasons, she found group work difficult and preferred working on her own both in the class and at home. The teacher Debbie also spent additional time outside class, supporting Cheri's writing, as she understood the importance of writing in Cheri's life and for her confidence in learning. Debbie's use of additional unrecognised, unallocated time meant that both Susan and Cheri were able to participate in learning, but raises issues of the extra resources offered by the teacher to sustain learners' motivation and progress.

Example 7 – Individual needs while working in a group

Susan attends the spelling class in her local drop-in centre. She is 70 and describes herself as 'quite deaf'. She wears hearing aids and can lip-read. Because her hearing impairment was not noticed at school, she did not understand many of her lessons and finds spelling particularly difficult. She wants to improve her own skills to be able to help her great-grandchildren who live close by. She has many interests and hobbies including local history and she is learning to use a computer. She felt embarrassed that her spelling let her down when she enrolled on a 'Computers Don't Bite' course. She was nervous about joining a spelling class as she felt she might be the 'worst person' there. She feared that she would face the same isolation and ridicule as at school.

Susan sits near the front of the class so that she can see the flipchart and lip-read what Debbie, the tutor, and other class members say. Debbie mainly teaches at the front of the class, involving everyone, using the flipchart to show examples. She explains, asks questions and encourages learners to use the flipchart to provide answers to the questions she writes. Much of Debbie's teaching involves group discussion, which includes asking for examples of how people might apply what they are learning. After discussion learners use worksheets, individually or together, to practise what they have learned. There is a typed list of ten to twelve questions with gaps to be filled in by the learners. When most people have finished, the group answers the questions out loud so people can fill in any gaps on their sheets.

Debbie used this method to teach prefixes. She explained the meaning of the terms and demonstrated on the flipchart. She then wrote a list of words including 'done', 'fulfilled' and 'expected' and the group discussed how they could be changed by placing 'un' in front of them. Most people contributed, adding new words to the list and as a group discussed making sentences with these words. The same method was used for working on suffixes. Susan found the group work difficult, as she could not easily hear what everyone was saying. The worksheet was less challenging, as she could ask Debbie questions to help her to fill it in. What she had not completed in class she took home to finish.

Debbie was aware that Susan found hearing in a group setting difficult and made individual time for her at the beginning and end of the class. She checked that she had understood and felt comfortable in the learning environment. With this support Susan completed her work at home, in her own time, and often did additional writing. Debbie enthusiastically supported this writing, recognising that this provided Susan with extra confidence and a bridge from a safe home environment to a more difficult classroom one. Often, when Susan handed her work in for Debbie to check, she added a note to tell her how she was feeling about learning. For example:

> I didn't understand this before but I do now.

Debbie was able to respond to this, checking what Susan was learning and how she was feeling about the learning process.

While Debbie mainly used a whole group teaching approach, she had found ways of engaging individually to create a supportive learning environment. Aware of Susan's fear of being exposed as a poor speller, she found ways of working with her that added to her pleasure as well as her skills. Susan summed up her learning as follows:

> I think English is fascinating and it's a subject you can go on and on with. The more you find out about it the more interesting it becomes. I think that's the way learning should be, isn't it?

Example 8 – Finding safe routes through learning

Cheri is 33 years old and was in the same class as Susan. She has bipolar disorder, which prevents her from working, as she is unable to stabilise her mood swings. She is often depressed and has become increasingly socially isolated. Cheri joined the spelling class at the drop-in centre to 'be able to use my brain'. This was especially important for her sense of self-worth and self-identity. It also enabled her to get out of the house and meet people – something she would not have been able to do otherwise.

Debbie, the tutor, was able to support Cheri in the class as she knew about her illness and some of the difficulties she experienced. For example, while her medication was being balanced, Cheri experienced paranoia, as well as difficulty relating to others. Debbie used time before and after the class to ask her how she was feeling. She also talked through additional writing and poetry that would fill the gap for Cheri until the next class. Cheri had done reasonably well at school and saw herself as someone who was coming to 'remember' rather than learn for the first time. She had lost some memory functions after ECT treatment and as a result of her continuing medication.

As she was able to complete the tasks in class she asked Debbie for additional work for 'the challenge'. Debbie found that although Cheri was a very competent reader she was unable to read books, as this created flashbacks to when she was hospitalised. Instead, Debbie worked on writing and developed her poetry-writing skills. Writing poetry enabled her to explore new forms of expression about her experiences. She became known as the class poet and had a poem published in one of the drop-in centre's booklets.

Cheri really wanted to do maths but she enjoyed the spelling class, particularly when she remembered things that she thought she had forgotten. This gave her confidence in her ability to learn while managing her illness. Her relationship with Debbie was particularly important as it showed that she could learn new skills and knowledge but also that she would be valued as a person while managing her illness. Debbie's understanding of the illness enabled Cheri to succeed in the spelling class. It also helped Cheri to find new forms of expression through her writing and gave her self-confidence to see herself as a learner.

While Cheri sometimes found it difficult to attend classes she felt that she always benefited from being there. The benefit extended beyond the knowledge of spelling that she was learning. It was about being accepted and valued as a person managing an illness. This acceptance and support enabled Cheri to enrol on and successfully complete a maths GCSE course at the same drop-in centre.

PRINCIPLE 5: Locate literacy learning in other forms of meaning-making. Recognise oral, visual, individual and group ways of communicating.

Language, literacy and numeracy are not always based on reading and writing. People often find, receive and share information by talking to others. Many ESOL learners may use one language for speaking and another for writing. People also make sense of the world by what they see and read in the street – such as bus numbers or road signs. They are constantly using and learning language, literacy and numeracy, in some way, as part of daily life. Sometimes people learn from each other and sometimes they teach each other at home, at work, in the community or in a class. Learning about language, literacy and numeracy occurs in many places and in many ways.

Focusing on people's existing oral and visual skills and interests encourages us to think about the ways that we communicate and make sense of the world, and how this can be incorporated in learning. Acknowledging people's understanding provides a strong foundation for developing new knowledge and skills. Acknowledging communication and learning outside the classroom makes it easier for learners to be confident about what is being 'added on' inside the classroom. For example, tutors who incorporate oral and visual methods in the classroom also provide the potential for more enjoyable, informal and creative learning.

People make sense of the world, both on their own and with others. For example, individuals find out information on a hobby like gardening, or use computer chat rooms as a social activity. Some people like working in a classroom group, as learning from and with others makes them feel more confident. Others, like Jason, from the maths class referred to earlier, prefer to work mainly on their own. Although Jason did enjoy joining in with group activities, he liked being given the choice to participate when he felt socially confident, without feeling under pressure to do so.

Where teachers take account of how adults learn both inside and outside the classroom, learners feel valued and enjoy learning.

Encouraging talking, looking and seeing also helps to develop the social aspects of learning in class. The two examples that follow show how oral, visual and communicative skills used in everyday life can be drawn upon in teaching and learning language, literacy and numeracy. The first shows how Amna, a bilingual literacy learner, used stories and games to develop speaking and listening skills for herself and for her children. She developed an understanding of how she was being taught, then applied these approaches to help her children learn: she became both learner and teacher.

In the second example, individual experiences of Somali house building methods were the basis of a group project. Speaking, listening and working within a group helped to develop confidence and new language skills as well as share cultural and practical knowledge and its meanings. Amna and the women in the Somali House project were able to move between teaching and learning, and one reinforced the other.

Example 9 – Developing speaking and listening through stories and games

Amna attends the same Bilingual Family Learning Centre as Fatima (see Example 5). She is from Oman and wants to learn English to help her children at school. Her older children help her to learn English at home, although she often has to ask them: 'What are you saying? Explain it to me slowly.' Although she is a fluent Arabic speaker she attends Arabic school to learn how to write in the language and to teach her children. She does this by setting them quizzes in Arabic at home. Amna speaks her first language at home with her small children but has started to introduce English phrases and words that they would hear at the crèche or nursery. For example, she says, 'Sit down please, go to sleep now.' She knows it will be important for her children to understand such instructions when she is not there.

The methods used by teachers at the centre try to mirror those found both at home and school. They rely on visual, oral, individual and group ways of working. These include stories, puppets, nursery rhymes, games and class discussion as well as worksheets and writing/retrieval tasks. Many of the texts used in the classroom were storybooks, often bilingual and traditional, depicting oral stories told by students. They are designed to be discussed in class and also to be taken home to be used and enjoyed and explored by both parents and children. Importantly, the stories can be used by people with different language skills and do not presume any literacy skills among the adults. The stories can be told, sung or mimed at home as they are often accompanied by puppets and cut-out

©digitalskillet/Shutterstock

characters. In class, the stories provide a good way of discussing cultural meanings and differences in groups of different language speakers that does not rely on text proficiency.

Parents often take the stories home to look at with children with accompanying practical activities like literacy and numeracy games. Parents with no reading skills, or confidence in English language, can support both their own and their children's learning by using their own home language and creating meanings by the process of looking and talking. This reflects everyday cultural practices, with stories being shared in many ways and across different languages.

Amna takes a story home every week and reads it to her children. 'My children like it every time, they ask me when am I going to read a story.' Sometimes she covers the words and shows her children how to make sense by looking at the pictures – a method she herself uses to learn in class. Her daughter, in a reception class, is not yet reading but, 'She saw the picture and she understands the story from what they say.' She knows that her younger son does not understand in the same way, but is still learning as he looks and listens. Because the centre uses similar visual and oral methods of teaching found at home and school, many of the women become more confident at supporting their children's learning. Amna and her children talk every day about what they have learned at school, and at home they read stories, play games and cut out and paste pictures for pleasure.

Example 10 – The Somali House workshop: Collaborative ways of working

The Bilingual Family Learning Centre used learners' oral and visual skills and their cultural knowledge to present a Somali House workshop. The workshop was designed to provide language practice and develop confidence in the learners taking part, as learners explained a theme and skills which they knew about. A small replica Somali house was available, which had been built by a previous group of learners. The workshop explored how people build and live in these houses. In class, information about house-building techniques was discussed in a group of pre-entry learners, with people individually seeking additional information in the local community. The tutor used an alphabet picture book asking for Somali words, for example, house, camel, friend, bark and tea. The group also worked on Arabic, Farsi and Bengali equivalents.

The women discussed these questions in Somali, checking meanings and comparing their experiences, often using the house as a visual cue. People used pictures, photographs and a mixture of languages to discuss rural life in Somalia. This prompted discussion about the civil war and migration. The tutor, Inga, pointed out that they would need to learn the words in English to be able to answer questions in the workshop. The group began to compile a simple dictionary. This included words like *tuuloh*, meaning 'village', and *dhig(o)*, meaning 'stick(s)', and prompted discussion about the use of plurals.

In the process of explaining the construction and use of the house, learners realised that they would have to do more than simply translate Somali words into English. They would have to explain how materials were used in the building process and discuss what is important to the family economy. Learners volunteered this information, first in Somali but then English, using words like *skin* and *house*. One learner brought a photo of herself as a child outside her family house to show the scale more clearly. Others worked together on posters and drawings, using labels to help people 'read' and understand the information they were presenting. The learners later said that the experience of preparing and delivering the workshop had been stimulating and rewarding.

3 Challenges and implications

In this section we look at the challenges and implications of teaching that acknowledges literacy as a social practice. We are not suggesting replacing existing ways of teaching adult language, literacy or numeracy with this approach. Using social practice offers an underpinning approach which, blended with other methods, enriches and adds to current ways of teaching and to learning. It is particularly significant at a time of increased national commitment and funding for basic skills. There are more learners and teachers and greater diversity in the learner population than ever before. This approach has many advantages for teaching diverse groups of adult learners, but it also poses challenges to everyone teaching and supporting adult learners.

What does this approach bring?

Using a social practice approach in teaching starts from the experience and perspective of the learners, rather than assuming – as can sometimes happen – that people have to fit, or be fitted, into existing systems and cultures. Research and practitioners' experience show that by connecting what people know and use outside the classroom to what they learn inside, it is possible to achieve a 'closer fit', making the learning both relevant and useful. Where people see the value in and connect with what they learn, they are more engaged and motivated. For example, Dee was able to overturn her negative views about DIY, as she saw it could help her realise her dream of independence. Sarah was more enthusiastic about learning measurement when she was able to see how she could use her skills to make shelves for her children's bedroom.

When people feel that their interests, experiences, talents and skills are recognised in the classroom or other learning setting, this encourages them to use and practise new skills in a variety of settings outside the classroom. It is also vital that people who come to learn are not just seen as learners; they often also support the learning of others. Amna saw herself as both a learner and teacher – these two things were interlinked in her family. She was aware of not just 'what' she learned but 'how' she learned it, using this to support the learning of her young children. Many women at the Bilingual Centre were teaching their children to speak or write in Arabic, using techniques they used to learn themselves. They also taught their children cultural knowledge about their extended families and places of origin.

Most adults are aware of, or with help can quickly identify, how they learn best. For example, Jason knew he preferred to learn on his own but also liked being able to join in with group activities when he had the choice. Susan became aware of how she learned best in a group when she experienced difficulties with her hearing. For her this meant asking questions in class time, and finishing things at home. Many learners know the pace they want to work at, and recognise the areas they find difficult or are confident with. They know what they enjoy and can respond to. A social practice approach means that teachers look at and plan teaching on the basis of learners' views and understandings of how they learn from their everyday lives. Practitioner research and student inquiry offer two ways of finding this out. Kay discovered through her questionnaire how people learned and used numeracy in their daily lives and she was then able to use this knowledge effectively in the classroom.

Locating learning materials and teaching resources in people's experiences and interests makes learning and skills familiar and valuable. Fatima was highly motivated to use and develop her notebook. She carried it everywhere to give her both information and confidence. Because she decided which pages to fill in at the centre she helped to identify the knowledge and information that was most useful for her everyday life. The notebook enabled her to learn in the classroom and to participate more independently outside of the house. Jack used authentic materials to learn, enabling him to maintain his interest in improving over several years of attending classes. By linking two worlds, and two identities as a farmer and a learner, he developed both his skills and his interests.

For many people, connecting their everyday world to learning meant they discovered new things, which became part of their widening interests and developing knowledge. Although most people were able to explain clearly what they wanted from a particular class – to learn spelling, or in Abdul's case grammar – they also enjoyed new and unexpected pleasures. Susan wanted to learn to spell but by attending the class became increasingly interested in and excited about the roots of language. She started using a dictionary for pleasure and explained that she had a different relationship to language and learning since coming to class, which exceeded her original aim of learning to spell. Cheri also discovered that she enjoyed writing poetry and felt this added an important dimension, not just to managing her illness, but to how she saw herself. Abdul was very clear that he wanted to learn grammar, because he enjoyed the mental stimulation and recognised this way of learning. Even though Duncan introduced the adventure story in a very low-key way, Abdul enjoyed the text as a story. He found that it was possible to learn from something that was imaginative and enjoyable – providing a different way of learning from his previous experience.

Acknowledging that lives may be complex and sometimes difficult, the social practice approach emphasises the need for teachers to listen and talk to find out what is important or significant for each person. Many people bring past or current experiences that affect how they are able to learn. For many adults, therefore, a sense of safety is a pre-condition for learning. This may be because of mental or physical health, or experience of violence, physical threat, intimidation, uncertainty and precariousness in their lives. These experiences, and the feelings associated with them, do not simply stay outside the classroom door. Where teachers are able to understand what the issues are for learners, they are more able to create a safe environment for learning. Debbie, as the teacher, was able to talk to both Susan and Cheri about what was difficult for them, from their everyday experiences, and discuss what support they needed in class. This was a mixture of 'looking' and checking in class and finding ways of 'hearing' how they were managing the learning within their life. Although Debbie was teaching spelling she understood that coming to class to learn was more than acquiring a skill in spelling – it was about working with a whole person. Dee's teachers were also aware that what was going on in her life could not simply be ignored and tried to find ways of working with this.

This approach does not offer a ready-made template for teaching, as it recognises that each learning environment and each tutor is different. Instead, it promotes the value of practitioner and learner inquiry. Rather than seeking the answer 'out there', this approach suggests that learners are able to ask questions and find answers themselves. This strengthens individual and collective practitioner knowledge and confidence. Teachers have professional knowledge they bring to practice, which can be applied to practitioner inquiry. Teachers make policy work through their practice in the classroom. They are therefore in a good position to know the significant questions that need to be asked or the issues that need to be explored. Kay and Debbie both carried out a small piece of practitioner research, which benefited their practice and fed into their college (see Ivanič, 2004). The approach also values learners as participants of the inquiry process. By connecting everyday lives to learning, it recognises that people ask questions and find answers in their daily lives. They develop a range of skills that can be used in the classroom to extend the nature and scope of learning.

The approach is a reflective process that supports critical questioning about what we do and why we do it. At a time when teachers are routinely dealing with the requirements of the system (assessments, targets and testing), it provides a reflective space to think about how these benefit learners, or how they can be adapted to do so. Rather than striving to be more efficient at administering the system, it asks what works well for the learner and what does not. It asks, as Kay did in her maths class, how practice can be developed and improved. In this way reflective practice is an essential part of initial teacher training and continuing professional development.

It is important to reflect upon what is possible and what is not. This approach is not a gold standard that everyone should adhere to. Teachers can use and adapt it as necessary and as practicable. We are aware that there are many differences in learners and places where teaching occurs – from a city centre to remote rural communities, from fixed college courses to short, work-based sessions and from flexible community provision to prison education. The principles will apply in different ways, in different settings. The core belief in recognising the everyday practices of learners is what remains constant.

This approach brings:

- a connection between learning inside and outside the classroom, making a 'closer fit' between life and learning. This helps to motivate and engage learners;

- an awareness that people learn in their everyday lives, an understanding that can be developed and built upon in class;

- an understanding that using authentic materials that are relevant to learners' lives can trigger and develop their interest and confidence in learning;

- a recognition that creating a safe, whilst stimulating, environment is essential for many adults to be able to participate in and benefit from learning;

- a commitment to thinking about how people learn, a continuing process supported by research and inquiry.

What are the challenges and implications?

Finding out about learners' everyday lives is not straightforward. It takes time. There are ethical and practical issues to be taken into account. Finding ways of asking about people's lives requires tact and sensitivity, knowing when to continue conversations and when to stop. It requires a positive response to what people are doing, rather than concentrating on what they cannot do. It requires an open account of why information about people is helpful and how this information will be used and recorded. Many people routinely have personal information taken from them without knowing how or where it will be used. Anyone involved in teaching or research needs to be aware of this so finding out about learners' lives does not feel like another official information-gathering exercise. In addition, there should always be absolute clarity between participants in a discussion which enables learners to say they do not want to participate.

Gathering information about people's lives

It may not always be appropriate to ask direct questions about learners' lives. For example, refugees and asylum seekers may be unwilling to talk about past experiences that may be painful or difficult, or still make them vulnerable. Women who have experienced violence, intimidation or trauma may find talking about some aspects of their lives too difficult, or may be unwilling to talk to someone they feel may judge them. Young offenders and learners in prisons may also not wish to talk about past or present circumstances and there are issues of recording confidential 'evidence' that may be admissible in a court of law. It is important, particularly in difficult circumstances, to provide positive signposts that lead to the future as well as find out about current and past experiences that may be difficult or negative. More generally, many people may prefer to talk about themselves as trust develops, rather than at the outset of their engagement in learning. Sensitivity about timing is critical to developing trust.

Practitioner research and student inquiry provide up-to-date and relevant information that supports reflective practice. However, there is a tension between generating information and people feeling that they are learning what they have come for. As we suggest above, learners know why they come and how they wish to learn. What is required is a careful process of negotiation to introduce new ideas that extend learning, through inquiry, while ensuring that learners progress and achieve in ways which meet their purposes. Teachers may feel there is not enough time to carry out additional tasks, and may be unwilling to become involved in something they view as slowing the process of learning literacy, language and numeracy. Practitioner research and inquiry should be part of initial teacher training so people are able to discover and appreciate its value to practice. It should also underpin continuing professional development, breaking down the mistaken belief that it belongs to 'experts' rather than being an essential tool for practice.

Using and developing materials

Taking account of learners' lives in teaching has implications for the range of resources used in the classroom. We suggest that materials and resources used inside the classroom should connect with people's interests, goals and passions outside of it. Does taking account of individual lives mean producing individual materials for each learner, which also have to be mapped onto the curriculum? This is clearly not possible, as there is not sufficient time or resources to do this. What it does mean, and this is the challenge, is that materials and resources should be used in ways that relate to each learner's life. This is as important as the materials themselves: good materials badly used do not produce good learning. There is a challenge, therefore, to adapt existing resources and develop new materials to enable learning to be relevant and useful in daily life.

We suggest two possibilities for the use of existing materials and resources. Of course there are others. The first is 'tailoring' and the second is 'translating'. Tailoring is using existing materials but linking them closely with the individual's life so that the subject content applies to their everyday experience. Jack's learning is a good example of this. His support tutor talked to him and listened carefully to what was going on in his life. With his agreement, this then became the basis for tailoring the materials and resources she used. She changed the vocabulary he was learning, rejecting those words that were not relevant, and she used Jack's own words as the basis of his computer work. Much of the subject content of Jack's learning was tailored to his identity and interests around farming.

Translating is applying experience or understanding from individual life to more general materials so that learners are able to translate their previous experience to the task in hand. Kay's use of the maths worksheets is a good example of translating the skills that Sarah wanted to learn in her everyday life into teaching resources about measurement, which Sarah was able to understand and use. The same applied to her work with Katrina and the nappy-changing rotas. Kay translated multiplication tables into an everyday setting and then back into a maths worksheet.

Developing new materials presents many challenges, often connected to our own levels of skills and openness to learn as professionals. There are many exciting and creative ways of introducing language, literacy and numeracy into the classroom, including websites, graphics, digital images, mobile phones, video, blogging and creative or free writing. All of these can be included within the structure of the curriculum, yet we miss many opportunities to try out new approaches. Often learners have the skills and imagination to develop new materials and resources. While time, or lack of it, is always a pressure, working with colleagues or learners to explore new materials and resources is an important part of practitioner inquiry and the continuing process of learning about learning. The Somali House project provided creative ways of developing new resources and materials. Some were almost completely learner-led and some were developed collaboratively with the teachers. This collaboration benefited all involved, including those who came to the workshop. The women learners learned about developing materials and resources through participation and understood how they could be used for transferring information and knowledge to others.

A whole people approach

A social practice approach acknowledges the importance of the social and emotional aspects of teaching and learning, recognising learners as whole people, with lives outside the classroom. It sometimes involves dealing with difficult issues. As we have shown, these could relate to physical or mental health, isolation or poverty (see Bynner and Parsons, 2006).

Responding to these issues in a classroom requires people management and personal skills as well as subject knowledge. Where do teachers best learn these skills? Many practitioners we talked to felt unprepared by initial teacher training to know how to respond to some of these issues. Many said they learned on the job, were helped by an experienced colleague, or through trial and error. When dealing with the complexity of some learners' lives these approaches do not seem on their own to be adequate. They need to be integrated more thoroughly in training and continuing professional development.

There are tensions about what is professionally appropriate, in creating and maintaining boundaries and taking care of individuals while being responsible for the learning of many. These tensions are not easy to resolve. Initial training should address these aspects of adult teaching and learning, giving teachers the insight and confidence they need to work effectively with people in the classroom. The use of coaches and mentors is a good way of providing support for teachers who sometimes feel isolated – particularly if they are part-time or work in rural or work-based provision (see Derrick and Dicks, 2005). These issues need to be made more visible, not ignored or papered over, in thinking about how adults learn and how best to support their learning.

A social practice approach is based on a process of 'talking and listening'. An implication of using this approach is to find out what people want from learning. However, this also needs to include 'doing and achieving' within a curriculum and assessment framework. Listening to what learners want and fitting this within the curriculum can potentially feel like a tension. People come to learning with a variety of life experiences and therefore have differing expectations of what they want to achieve. Where learning is organised around the literacy, numeracy and ESOL core curricula, there are often assessments or tests to measure how much learning has taken place. For many learners this is positive, as they want to measure their own performance or to see how far they have gone on their own learning journey (see Ward and Edwards, 2002, and also Davies, 2005). For others this is less positive as they do not want to take a test in case they fail, become too anxious, or feel the test has no relevance to their lives and goals. For example, Susan did not want to take the national test as she felt that, at 70, it was not important to her. Jack also did not feel he needed this external measurement of his learning. On the other hand, Abdul was keen to take the national test, and do well, for his sense of self-worth and to help him find a job. Dee also felt that she would benefit by having a national qualification, something she had not achieved before.

Teachers often describe learners as having 'spiky profiles' to account for uneven levels of ability across different areas; for example, where someone may be good at reading and have knowledge and skills in other areas, but may find writing or maths difficult. This is often mapped against the 'curriculum offer' and where progress in the area of specific difficulty is measured. Although this is positive, as it recognises that people come with existing knowledge as well as new things to learn, it does mean that success is mainly measured by tests. Listening to learners shows this is not always positive or relevant to their lives. Tests are clearly a good thing for some people – learners tell us this. The challenge is that, once we hear what learners say, we need to find wider, additional ways of recognising, giving feedback and acknowledging learning progress, both inside and outside the classroom. Cheri's poems and the work around the Somali House workshop are good examples of non-assessed learning that existed beyond what was tested, which added to the learning process. Many teachers and institutions already find creative ways of giving 'in-house' certificates to acknowledge this learning. Increasingly, this is also being recognised more widely, resulting in national materials being offered to acknowledge the range of learning achieved (DfES, 2003).

Professional tensions

Working flexibly, creatively and responsively with individuals in a system of targets and assessment-driven curricula can create challenges in responding to the different needs of both. In adult literacy, numeracy and ESOL teaching we have called this a tension between two sorts of teaching professionalisms (see Ivanič *et al.*, 2006). One is rooted in being responsive to learners as individuals and is informed by social justice principles. The other is based upon being proficient in working with the core curriculum and the targets and assessments in the Skills for Life strategy. Clearly, both types of professionalisms are better blended to support learners individually and to maximise the opportunities created by the national emphasis on improving basic skills. It is important that both are introduced and supported through initial training and continuing professional development.

Implications and challenges of this approach

- Finding out information about learners' lives is not straightforward; it takes time and needs to be approached carefully and ethically, enabling learners to fully understand and have a choice in the process.

- Practitioner research and learner inquiry needs to achieve a careful balance between asking questions, reflection and achieving what learners come to learn.

- Materials need to be developed and used in a variety of ways. This may require finding new information and in some cases learning new skills.

- Recognising and supporting the social and emotional aspects of people's everyday lives in learning needs training and continuing support in people skills and classroom management.

- People come to learn for a variety of reasons. To be able to match what they want with what is on offer requires talking and listening carefully to learners, which take time and resources.

In conclusion

We have presented a social practice approach to teaching adult literacy, numeracy and ESOL in this guide, which draws on current theory and research. We believe that the principles of social practice pedagogy apply to most teaching, and underpin good practice generally, but are still underdeveloped in the field of literacy. To show the principles in action we have used examples from research. These have been commented on by a range of experienced practitioners and teacher trainers. As we suggested at the beginning, this is not a simple recipe book for successful adult teaching – it is to support reflection and inquiry in the interests of learners' motivation, engagement and pursuit of their own purposes.

In initial teacher training, some practice approaches can develop discussion about who the people are that come to learn, why they come and what they bring from their everyday lives. By exploring the connection between lives and learning that is at the heart of this approach teachers can see and value the skills, knowledge and interests that learners bring with them. Understanding this connection provides ways of working with people to make a better 'fit' between what they learn in the classroom and what they learn and use at work, in families, social networks and communities. This approach requires thinking about what language, literacy and numeracy are within training and in continuing development. It questions a narrowly defined skills agenda, which is part of much basic skills learning. It provides an important alternative view to this, asking what literacy is and how this needs to be accounted for in literacy learning. Looking at the additional resources provided at the end of this guide can develop this understanding further.

Each example shows a real person who has a past, individual present circumstances and an imagined future. The examples can be used to discuss ways of working with people that take account of lives outside the classroom, looking at potential strategies and constraints. They can be used to stimulate discussion and connect to practitioners' experiences of learning new skills and using language, literacy and numeracy in their own lives – learning about learning. By highlighting the importance of practitioner research and inquiry, the approach emphasises teachers asking questions themselves. In training, this can be done collaboratively or individually, to support critical reflection.

For some teachers, particularly those working on their own, social practice can support a reflective process connecting them with other ideas and materials outside of their own practice. Language, literacy and numeracy teachers are often working in situations where their specialist topics are delivered alongside other subject work, for example in vocational subjects such as hairdressing or engineering. This learning needs to be well embedded in order for students to see the relevance for work and life and to achieve their vocational and language, literacy and numeracy qualification. Although effective embedding depends on factors such as staff values and beliefs, teamwork and organisational factors, it is enhanced by a literacy as social practice approach that ensures literacy is not just taught as a decontexualised skill.

It may be helpful for individual teachers to reflect on how they feel able to work and how they would like to work. Practice is set within educational structures and systems; sometimes these present opportunities, sometimes they present constraints, both real and felt, and frequently they create a tension between the two. As one teacher said, 'My learners are recovering from mental ill-health and we design and plan literacy around their interests but we are under pressure to put them through the national test.' Is it possible to juggle what might feel like two sets of competing interests, those of the learner and those of national assessment criteria, and how does a social practice approach help?

This approach enables us to understand what the tensions are between lives and learning and between teaching and the curriculum. It does not provide simple solutions to difficult tensions and issues but it can support realistic reflection on what is possible, practical and positive to aim for in teaching adult language, literacy and numeracy, and the significance of this in the lives of those who come to learn. Importantly, it requires thinking of literacy not simply as a set of skills that can be taught in a disembodied way in the classroom, but as part of people's everyday lives and experiences.

References

Appleby, Y. and Hamilton, M. (2006) 'Literacy as social practice: Travelling between everyday and other forms of learning', in P. Sutherland (ed.) *Lifelong Learning: Contexts and Concepts*. London: Routledge.

Barton, D. and Hamilton, M. (1998) Local Literacies: *Reading and Writing in One Community*. London: Routledge. This was one of the first books to look in detail at how people use language, literacy and numeracy in their everyday lives. It provides vignettes about people's uses of, and passions relating to, literacy, as well as developing social practice theory from these lives.

Barton, D., Appleby, Y., Hodge, R. Tusting, K. and Ivanič, R. (2006) *Relating Lives and Learning: Adults' Participation and Engagement in a Variety of Settings*. London: NRDC. This report from the Adult Learners' Lives project describes the second phase of research, which looked at learning in community settings. It is available at **http://www.nrdc.org.uk**

Baynham, M., Roberts, C., Cooke, M., Simpson, J., Ananiadou, K., Callaghan, J., McGoldrick, J. and Wallace, C. (2007) Effective teaching and learning: ESOL (London: NRDC). It is available at **http://www.nrdc.org.uk**

Bynner, J. and Parsons, S. (2006) *New light in Literacy and Numeracy*. London: NRDC.

Coben, D., Brown, M., Rhodes, V., Swain, J., Ananiadou, K., Brown, P., Ashton, J., Holder, D., Lowe, S., Magee, C., Nieduszynska, S. and Storey, V. (2007) Effective teaching and learning: Numeracy (London: NRDC). It is available at **http://www.nrdc.org.uk**

Davies, P. (2005) *Study of the Impact of the Skills for Life Learning Infrastructure on Learners. Interim Report on the Qualitative Strand*. London: NRDC.

Derrick, J. and Dicks, J. (2005) *Teaching Practice and Mentoring: The Key to Effective Literacy, Language and Numeracy Teacher Training*. Leicester: NIACE. This book provides useful information and strategies for teacher training and mentoring.

DfEE (1999) *A Fresh Start: Improving Literacy and Numeracy*. The Report of a working group chaired by Sir Claus Moser. London: Department for Education and Employment.

DfES (2003) *Planning Learning and Recording Progress and Achievement: A Guide for Practitioners.* This guide offers a three-layered description of a cycle of learning, which can be negotiated between learner and teacher. It covers setting goals, planning activities, reviewing progress, assessment, review and further progression. It is available at **http://www.dfes.gov.uk/readwriteplus**

DfES (2006) *Personalising Further Education: Developing a Vision.* London: Department for Education and Skills.

Hamilton, M. and Hillier, Y. (2006) *Changing Faces of Adult Literacy, Language and Numeracy: A Critical History.* Stoke on Trent: Trentham Books. This book examines key moments in the history of adult language, literacy and numeracy, identifying levers of change. It provides the context and critical history to guide practitioners working towards national qualifications as well as researchers and policy makers in the field.

Illeris, K. (2002) *The Three Dimensions of Learning: Contemporary Learning Theory in the Tension Field between the Cognitive, the Emotional and the Social.* Roskilde University Press (distributed by NIACE). Illeris provides a useful overview of the main learning theories and argues that to be useful in understanding learning we need a framework that includes cognitive, emotional and social aspects.

Ivanič, R. (ed.) (2004) *Listening to Learners: Practitioner Research on the Adult Learners' Lives Project.* London: NRDC. This short report describes six practitioner research projects that were part of the Adult Learners' Lives project. It can be downloaded from **http://www.nrdc.org.uk**

Ivanič, R., Appleby, Y., Hodge, R., Tusting, K. and Barton, D. (2006) *Linking Learning and Everyday Life: A Social Perspective on Adult Language, Literacy and Numeracy Classes.* London: NRDC. This report from the Adult Learners' Lives project describes the classroom phase of the research and its findings. It is available at **http://www.nrdc.org.uk**

Tusting, K. and Barton, D. (2003) *Models of Adult Learning: A Literature Review.* London: NRDC. This literature review provides a comprehensive and accessible overview of the main models of adult learning relevant to basic skills. The summary report is available at **http://www.nrdc.org.uk**. The full report is now published by NIACE (2006).

Ward, J. and Edwards, J. (2002) *Learning Journeys: Learners' Views on Progress and Achievement in Literacy and Numeracy.* London: Learning and Skills Development Agency. This accessible report describes an action research project that looked at adult learners' experiences. It used a journey metaphor to explore what progress meant to adult learners and what affected some of the differences people described.

Further reading

Books

Cope, B. and Kalantizis, M. (eds) (2000) *Multiliteracies: Literacy Learning and the Design of Social Futures.* London: Routledge. This edited collection provides an introduction to the Multiliteracies approach. This work develops the body of work, referred to as New Literacy Studies, which is critical of literacy as a simple functional skill.

Fowler, E. and Mace, J. (2005) *Outside the Classroom: Researching Literacy with Adult Learners.* Leicester: NIACE. This book presents both theory and practice of a social perspective of adult language, literacy and numeracy. It provides a useful overview of a social practice approach and practitioner examples of how this approach has been applied.

Horsman, J. (2000) *Too Scared to Learn: Women, Violence and Education.* London: Lawrence Erlbaum. This book provides an accessible introduction to the issues of violence and learning. It discusses working with the effects of violence and trauma through working with the whole person to break the silence that often surrounds these experiences, both for individuals and in educational programmes.

Learning Connections (2004) *An Adult Literacy and Numeracy Curriculum Framework for Scotland.* Edinburgh: Communities Scotland. The Scottish Framework provides a national model based upon a social practice perspective. It has a community wide vision of language, literacy and numeracy learning. It is available at
http://www.lc.communitiesscotland.gov.uk

Pahl, K. and Rowsell, J. (2005) *Literacy and Education: Understanding the New Literacy Studies in the Classroom.* London: Paul Chapman. This book explores the implications of using a social practice approach in school classrooms. Although adults are not the focus there are some useful insights that are relevant to adult classrooms and links to home literacies.

Papen, U. (2005) *Adult Literacy as Social Practice: More than Skills.* London: Routledge. This book looks at adult literacy as social practice. Written to support a distance learning MA, it provides a good overview of relevant theories and debates. It provides readings to guide reflection, discussion and debate.

Pitt, K. (2005) *Debates in ESOL: Teaching and Learning. Culture, Communities and Classrooms*. London: Routledge. This book, written to support a distance learning MA, provides a clear and accessible overview of some of the issues in ESOL teaching. It contains useful readings and discussion points.

Journals

Reflect. NRDC journal with useful information and comment from policy to practice at **http://www.nrdc.org.uk**

Literacy and Numeracy Studies: An International Journal in the Education and Training of Adults. This is a useful international journal covering literacy and numeracy.

Resources and websites

These are the main sites to access information from, which will also provide links to smaller sites. Many smaller sites themselves change, become obsolete or are not updated frequently.

Australian Council for Adult Literacy	http://www.acal.edu.au
Basic Skills Agency (BSA)	http://www.basic-skills.co.uk
BSA in Wales	http://www.basic-skills-observatory-wales.org
Communities Scotland	http://www.lc.communitiesscotland.gov.uk
Lancaster Literacy Research Centre (LLRC)	http://www.literacy.lancaster.ac.uk
National Adult Literacy Agency (NALA) (Ireland)	http://www.nala.ie
National Institute of Adult Continuing Education (NIACE)	http://www.niace.org.uk
National Literacy Trust	http://www.literacytrust.org.uk
National Research and Development Centre for Adult Literacy, Numeracy and ESOL (NRDC)	http://www.nrdc.org.uk
National Center for the Study of Adult Learning and Literacy (NCSALL) (US)	http://www.ncsall.net
Network for Workplace Language, Literacy and Numeracy	http://www.thenetwork.co.uk
Read, Write, Plus (DfES)	http://www.dfes.gov.uk/readwriteplus
Research and Practice in Adult Literacy (RaPAL)	http://www.rapal.org.uk
RipAL (Canadian practice and research network)	http://www.nald.ca
Talent	http://www.talent.ac.uk